Patios, Driveways, and Plazas

Coca-Cola
3 FT 3 IN
4 FT 3 IN

Courtesy of PaverModule

Patios, Driveways, and Plazas

The Pattern Language of Concrete Pavers

David R. Smith

The Interlocking Concrete Pavement Institute

4880 Lower Valley Road, Atglen, PA 19310 USA

Library of Congress Cataloging-in-Publication Data

Smith, David R. (David Randolph), 1953-
Patios, driveways, and plazas : the pattern language of concrete pavers / David R. Smith.
p. cm.
1. Pavements -- Terminology. I. Title.
TE250 .S63 2000
625.8'4--dc21
2001005738

Designed by Bonnie M. Hensley
Cover design by Bruce M. Waters
Type set in Exotc350 DmBd BT/Zurich Cn BT

ISBN: 0-7643-1561-7
Printed in China

Published by Schiffer Publishing Ltd.
4880 Lower Valley Road
Atglen, PA 19310
Phone: (610) 593-1777; Fax: (610) 593-2002
E-mail: Schifferbk@aol.com
Please visit our web site catalog at **www.schifferbooks.com**
We are always looking for people to write books on new and related subjects. If you have an idea for a book, please contact us at the above address.

This book may be purchased from the publisher.
Include $3.95 for shipping.
Please try your bookstore first.
You may write for a free catalog.

In Europe, Schiffer books are distributed by
Bushwood Books
6 Marksbury Ave.
Kew Gardens
Surrey TW9 4JF England
Phone: 44 (0)208 392-8585
Fax: 44 (0)208 392-9876
E-mail: Bushwd@aol.com
Free postage in the UK. Europe: air mail at cost.

Contents

Acknowledgments

Italian restaurants can have a special effect on people. The idea for this book was hatched in one with Tina Skinner from Schiffer Publishing Group. In the not too distant past, she skillfully brought forth two idea books on upscale paving for homeowners called *Creative Patios* and *Walkways and Drives: Design Ideas for Grand Entrances*.

This book was born to give ideas and more. It was created to show the substance and structure of design with concrete pavers. It's both an idea book for homeowners, plus a resource for those who design for a living. Tina Skinner and Peter Schiffer deserve thanks for seeing its potential.

Finding the recipe for a book and preparing one doesn't happen instantly. Both take time. A debt of appreciation goes to the Marketing Committee and Board of Directors of the Interlocking Concrete Pavement Institute. They allowed me to take the time to think about and prepare this book. Over the years they provided many trips (mostly for purposes other than picture taking) where I photographed some of the pictures credited to ICPI. In addition, there are many pictures from stories that appeared in the *Interlocking Concrete Pavement Magazine*. Again, this book would not have been possible without support from the concrete paver industry association.

The right ingredients prepared and presented well are Italian restaurant essentials. The salad, pasta, sauce, spices, cheese, wine, and dessert combine to leave a lasting impression. The best paver projects in North America have come from ICPI members' kitchens. This book would not be possible without their generous contribution of photographs. It is a delight to present their work and I trust it will impress and inspire readers.

This book has been incubating since the early 1990s. Many of the ideas in it were inspired by Gary E. Day, AIA Associate Professor of Architecture at the State University of New York in Buffalo, and by Anton Harfmann, Professor of Architecture at the University of Cincinnati, Ohio. They developed a curriculum for teaching architecture and landscape architecture students the design language of concrete pavers. This book continues their ideas and a great deal of thanks is due to them. I hope they and others can use this book in their teaching endeavors, design practice, and construction businesses. The least I can do to repay Gary and Anton is by taking them to an Italian restaurant. You never know what good ideas might result from it.

Courtesy of Barkman Concrete

Introduction

The development of vocabulary and sentences result in the articulation of moods, ideas, and concepts. In written and verbal forms, they support language, culture, and life. This book borrows language basics to help explain the design vocabulary of concrete pavers. Just as language gives power to create an infinite variety of sentences and conversations with corresponding character, mood, tone, and feeling, the pattern language of concrete pavers gives tools to create an infinite variety of messages and moods. Concrete pavers help communicate how a place works, and how to work and live in a place. The messages come from the size and configuration and the paved area, paver shapes, patterns, textures, and colors. These components combine in countless ways to make a design language.

The language of segmental paving is very old. Pavement made with tightly fitted units was first recorded by the Minoans some 7,000 years ago. About 2,000 years ago, cheap slave labor and military dominance enabled the Romans to build the first multi-national highway system with segmental pavement. When completed, it was longer than the interstate highway system in the United States.

Since recorded civilization, practically every culture used, and continues to pave with, segmental pavements. In North America, cobblestone streets and clay brick supported the early growth of major cities. Unfortunately, the memory of these types of segmental paving faded from 1920s to the 1970s. North America seems to be the only culture that experienced a lapse of memory on how to design with segmental units. This was due to

An ancient language and early precedent to concrete pavers: a part of the 2,000 year-old Roman international road system.

Concrete pavers: the modern version of an ancient way of building pavements. *Courtesy of PaverModule*

rising labor costs, the proliferation of the automobile and its need for a smooth surface, and an inexpensive petroleum by-product called asphalt that created smoother roads. Suddenly, the language of pavements in urban places became very boring.

The memory lapse of several generations of North American engineers and architects resulted in little working knowledge of the design vocabulary and technology of segmental paving during the mid-20th century. Existing knowledge and practice faded for lack of use and improvement.

Fortunately, the lapse has been temporary. A half-century obsessed with the growth of monolithic asphalt was a blip, a perturbation, in seven millennia of segmental paving throughout the world. The advent of interlocking concrete pavements in the 1970s revived segmental paving in North America. Concrete pavers re-introduced segmental paving in landscape and construction trades, higher education, the professions of urban design, architecture, landscape architecture, civil engineering, and in life's daily experiences.

Since the start of the North American concrete paver industry, millions of units are found in just about every place. The global industry makes 1,000 square feet (100 m^2) every second of every working day. The North American industry is quickly approaching annual sales of 500 million square feet. (50 million m^2). Users recognize that concrete pavers satisfy the need for a safe, smooth, durable and cost-effective riding surface for vehicles. More importantly, many recognize the elegance and human scale that concrete pavers contribute to the life and spirit of a place.

Whether used in residential, commercial, or municipal applications, concrete pavers fit many projects. Design professionals (practicing and student), landscape, hardscape, and paving contractors, and homeowners conceive these projects. This book is written for these readers. It is not a technical or installation guide, but a resource of design ideas for the early stage of projects. When information on specifications, construction, and maintenance are needed, they are available from the Interlocking Concrete Pavement Institute (www.icpi.org) or from its members. Many Institute members who contributed to this book are found in the Resource Guide in Chapter 10.

All languages evolve from influences in culture and technology. As time passes, language develops richer images and meanings, with finer semantic shades. Yet the basic structure still remains. The same holds true for concrete pavers. Shapes, patterns, textures, and colors continue to evolve with user needs and design trends of the culture. Yet concrete pavers remain the avenue to communicate the sense of a place, marking the path of human life and history.

David R. Smith
Interlocking Concrete Pavement Institute
Washington, D.C.

Chapter 1: Nouns

Stack Bond

Courtesy of Basalite

Pattern language fundamentals start with stack bond, a grid-like arrangement of pavers whose joint lines continue in perpendicular directions. Square and/or rectangular shaped units provide the pattern building blocks. They are arranged according to three aspects: the underlying grid or field, the color of the pavers, and frames to subdivide the fields (grids within a grid). Frames are defined by varying colors, paver sizes, and textures.

Nouns are places, persons, or things. Stack bond functions similarly by defining a place with its strong grid pattern. Colors inject personality and character. Frames subdivide the overall pattern, defining or highlighting site features. They are things such as tables, seating, fountains, pools, entrances, and buildings. This chapter shows many combinations of fields, colors, and frames to suggest formal (proper nouns) and informal places.

One Field and One Color

When one color dominates, it makes a formal presentation as in this backyard. *Courtesy of E. P. Henry Corporation*

The grid field of the stack bond and fountains communicate a formal place. However, their grids present an opportunity for playful freedom for children. *Courtesy of Paver Systems*

One Field with Different Colors

Varying the colors in this field is a delightful, unexpected way to transform this sidewalk. *Courtesy of Site Technologies*

Here are units of the same colors with different sizes arranged in two nearby fields in a shopping mall plaza. The larger units on the left side feel a bit more formal than the smaller ones on the right side. *Courtesy of Site Technologies*

A grid field is maintained, but the size and colors of the concrete pavers vary within it. *ICPI*

The field alignment with a contrasting color marks paths and edges around this pool. *Courtesy of Paver Systems*

Various colors in the pattern in this backyard patio relaxes the formality of stack bond. Contrasting colors at the steps help mark the edge of the pavement. *Courtesy of Barkman Concrete*

Seven colors in stack bond suggest waves at the waterfront location of Jack London Square in Oakland, California. *ICPI*

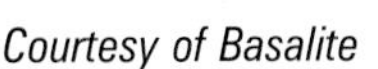

Courtesy of Basalite

Facing Lake Ontario, the plaza in front of the National Trade Centre in Toronto, Ontario, uses four paver colors arranged in a computer-generated, mathematical sequence to create wave-like patterns. The project was designed by Jerry Clapsaddle, one of his many supergraphic projects for public spaces using concrete pavers. *ICPI*

Distance and perspective allow the wave patterns to appear. *ICPI*

ICPI

Like nature, each wave is different. The pedestrian experiences motion while walking through a two-acre plaza of concrete pavers.
ICPI

Bending the stack bond pattern (it's almost running bond) with contrasting colors yields waves near the beach at Ft. Lauderdale, Florida. *ICPI*

Waves wash past this sign for surfers.
Courtesy of Paver Systems

Contrasting Colored Fields and Frames

Large frames define sitting areas around this swimming pool. The frame and fields within them use the same sized units but with different colors. *Courtesy of Paver Systems*

Contrasting Colors and Sizes for Fields and Frames

Small frames in small places imply less formality. *Courtesy of E. P. Henry Corporation*

A small frame of pavers around large fields of paving slabs with light-colored units provides just enough formality to this entrance and swimming pool. *Courtesy of PaverModule*

Courtesy of PaverModule

Slightly larger fields with the same sized frame grace this residence. *Courtesy of PaverModule*

Courtesy of PaverModule

Repetition of small and large fields sets the stage for this resort swimming pool. The grid begins with larger units in stack bond unit, and it is framed with smaller units of a different color. The pattern frames trees, sitting areas, and even a large chess board. *Courtesy of PaverModule*

Courtesy of PaverModule

Sometimes the frame can be large enough to hold its own pattern, further distinguishing the fields from one another. *Courtesy of PaverModule*

The diagonal stack bond with its own pattern frames the one-color field inside it. That field frames the rose in this award-winning design by Italian designer Giuseppe Abrancati. *Courtesy of Gappsi*

Like framing a picture, framing pavers with a different material makes them stand out from their surroundings. At the Toronto SkyDome stadium, the framing of stack bond paving units by another paving material breaks up a large space into smaller, more human scale areas. *ICPI*

The framed stack bond in this formal plaza at a shopping mall may have suggested arranging the pumpkins in a stack bond as well! *Courtesy of Site Technologies*

Similar paver colors, alignment with architectural building details, enclosed by other pavement materials reinforce the formality and dignity of this setting. *Courtesy of Mutual Materials*

Changing Field Direction

Combining a change in unit size and pattern (or grid) direction with use of several colors helps define zones of activity for pedestrians around this entrance. *Courtesy of Pavestone*

A stack bond pattern under a sitting area receives definition when framed or enclosed by another field pattern placed in a different direction. *Courtesy of E. P. Henry Corporation*

The structure and formality of stack bond can be relieved through opening joint spaces. These are filled with small-sized aggregate. *ICPI*

Subtracting pavers means adding spaces for joints. These are softened with grass, reducing stormwater runoff. *Courtesy of Nikko*

A field of larger paving slabs breathes with joints filled with grass. *Courtesy of Perfect Pavers*

Random subtraction of a few paving units from the pattern, plus the orientation of the textured surfaces add informality and intimacy to this small sitting place. *Courtesy of Nikko*

The one field starts disassembly with the emergence of a diagonal field merging into the pattern. *ICPI*

Ad Hoc Stack Bond

Pavers removed and placed aside in "vertical stack bond" near utility repairs instantly becomes a children's playground. *ICPI*

Chapter 2: Adjectives

Stacked Variations

Courtesy of Basalite

Adjectives depend on nouns, and modify them. The pattern variations in this chapter depend on the underlying grids of stack bond, the nouns. They also transform them. The transformation can be subtle or grand, depending on shapes, colors, and patterns.

Simple transformations are presented first and they move to more complex layouts. For example, the parquet pattern is a simple variation on stack bond. Rectangles stacked together are transformed to squares. At the complex end of the spectrum are squares, neatly stacked, but uniformly spaced apart. Units fill spaces between the squares that aren't square or rectangular. These infill pavers transform them. New patterns emerge from the combination, but they still depend on the neatly stacked squares.

Parquet

Stack bond is a grid of square pavers where joint lines run in perpendicular directions. Parquet patterns modify this theme by grouping a few rectangular pavers together to make squares. When assembled, joint lines between each square run continuously in both directions.

The color variations make this classic parquet pattern a bit more festive and less formal. *Courtesy of Pavestone*

Edison Field, a baseball stadium in Anaheim, California, uses parquet patterns grouped in contrasting colors at its entrance. *Courtesy of Davis Colors*

Contrast marks the curb ramp. *Courtesy of Bayer*

At Key West, Florida, a public plaza frames a parquet pattern with a contrasting color to make the immense area more human in scale. *Courtesy of PaverModule*

Courtesy of PaverModule

The patterns show the effect of framing one and two pavers around a small square. These are variations of the parquet pattern. *ICPI*

Courtesy of Pavestone

Randomness from order: A repeating pattern of stacked square units surrounded by rectangular ones (a "K-pattern") makes a random appearing field. *Courtesy of PaverModule*

Another example of the random appearing "K-pattern." *Courtesy of E. P. Henry Corporation*

The colors add to the randomness. *Courtesy of Unit Paving*

The variations on the following pages use square pavers arranged in a grid. However, the field of squares are spaced apart and pavers wrap around them. The pavers around the squares allow joint lines to be maintained in both directions. Colors and textures contrast the squares from the shapes around them. That contrast can make the squares dominate visually or the surrounding pavers can jump out. Squares and surrounding units can blend by being the same color, or nearly the same color.

Same Pattern, Change a Color

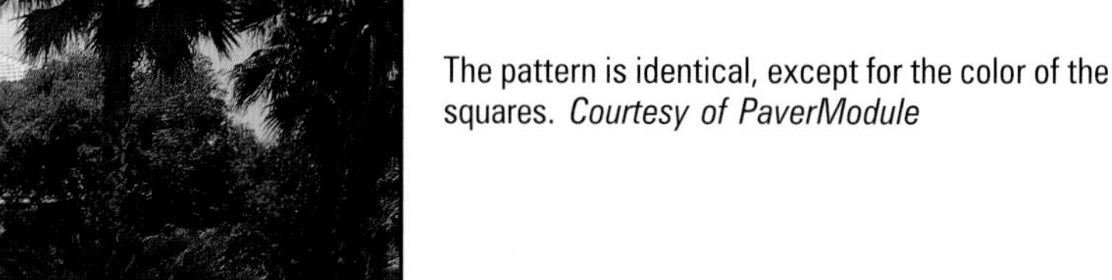

The pattern is identical, except for the color of the squares. *Courtesy of PaverModule*

Courtesy of PaverModule

An orderly arrangement of square units wrapped by smaller squares. *Courtesy of PaverModule*

This is almost the same arrangement of squares, with rectangular pavers surrounding each, but the strength of the random colors hides the pattern. *Courtesy of PaverModule*

Courtesy of PaverModule

ICPI

A shopping district in Boca Raton, Florida, marks a linear park with a variation of stack bond using different sized and colored units.

Courtesy of Paver Systems

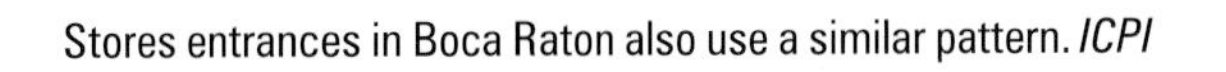

Stores entrances in Boca Raton also use a similar pattern. *ICPI*

ICPI

A unique use of two squares and rectangular pavers make this driveway almost tile-like. Joint lines run in one direction toward the garage. *Courtesy of Stone Products*

Joint lines run toward the house, but they take a delightful detour around the squares with the introduction of another pattern. *Courtesy of Paver Systems.*

Opposite page:
The squares speak out in this almost tile-like pattern. Note that the joints lines spin off the tips of the diamonds (squares turned 45 degrees) implying continuity. *Courtesy of PaverModule*

Sometimes the contrast in colors needn't be great, but just enough to announce the pattern. *Courtesy of E. P. Henry Corporation*

NO DIVING
6 FT.

The colors calm the pattern. *Courtesy of PaverModule*

Opposite page:
Here is the same pattern as the previous page except with hushed squares from the blended colors. *Courtesy of PaverModule*

Those squares do like to jump out and speak!

ICPI

ICPI

Courtesy of Paver Systems

Courtesy of E. P. Henry Corporation

Courtesy of E. P. Henry Corporation

Courtesy of Borgert Products

Courtesy of Pave Tech, Inc.

Running Bond

Courtesy of Basalite

When people think of old clay brick pavement, they often visualize running bond pattern, i.e., pavers set end-to-end, offset from the next course by a half a unit. Joint lines continue in one direction only, either going across the pavement or running through its length. A large number of brick streets in running bond still exist, and many are hidden under more recent paving materials.

The historic appearance of brick streets is easily replicated with concrete pavers, and with a smoother riding surface. Running bond with concrete pavers is occasionally used in streets, but it appears mostly in pedestrian applications and residential driveways. No matter where it's used, it speaks movement, action, and direction. It makes a path to or through a place.

Next to Ninth Street is Cleveland's famous Rock and Roll Hall of Fame and Museum. Six acres of concrete pavers by Lake Erie surround the museum, plus a nearby science museum and a football stadium. *ICPI*

Opposite page:
As a traditional old street pattern, Ninth Street by Lake Erie in Cleveland, Ohio reflects the pattern of the old cobblestone streets. A narrow grid of stack bond pavers in a contrasting color divides the vehicular traffic. *ICPI*

Given enough length, intersecting bond patterns direct the eye and pedestrian movement. Like the wake of a boat, the entrance walkway boldly pierces the plaza's wave pattern. *Courtesy of Tremron*

Opposite page:
A traditional laying pattern makes modern waves by changing colors. It supports the almost Jetsons-like character of a sports arena in Florida. Compare this to wave patterns created with stack bond in Chapter 1. *Courtesy of Tremron*

Color and bond direct a path by the Atlantic Ocean. The pattern pierces another. *Courtesy of Interlock Paving Systems*

Paths direct baseball fans to Turner Field in Atlanta, Georgia. *Courtesy of Site Technologies*

Watch like an Egyptian: all roads lead to the movies. *Courtesy of PaverModule*

Courtesy of PaverModule

Finish, Size, Color, and Pattern Direction

While these homes might be formal looking, pavers with a rough, tumbled finish, plus various colors and sizes suggest classic stone or worn brick. Larger units suggest more formality than smaller units. Running bond perpendicular to the direction of travel supports a formal setting. These subtle variations grace the following projects.

Courtesy of PaverModule

Courtesy of PaverModule

Courtesy of PaverModule

Courtesy of PaverModule

Stack bond borders mark the running bond, front entrance with units of varying size and color. *Courtesy of PaverModule*

The backyard catches golden rays of sunset, catching an altogether different feel than the front entrance on the previous page. *Courtesy of PaverModule*

Running bond flows with the drive and entrance. *Courtesy of Matt Stone*

The clean, continuous joint lines and subtle color blend maintain a certain dignity while alternating the unit size keeps the pavement from becoming too stuffy and intimidating. *Courtesy of R. I. Lampus Co.*

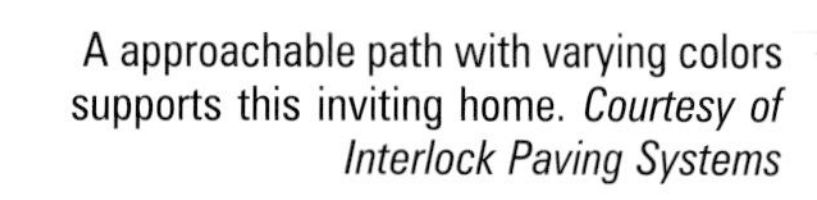

A approachable path with varying colors supports this inviting home. *Courtesy of Interlock Paving Systems*

Running bond easily adjusts to bends in a path. *Courtesy of Belgard-Permacon*

Running bond flows from one direction into another as it work its way up the steps. *Courtesy of Belgard-Permacon*

A throw rug always fills a space as in this award-winning design. The use and direction of running bond accentuate the stretching. *Courtesy of Gappsi*

The direction of the bond pattern enlarges the patio. *Courtesy of Lafarge Pavers and Walls*

The narrow path to this garden is lengthened by the direction of the bond. *Courtesy of Interlock Paving Systems*

These two examples from Old Quebec City indicate direction in pedestrian areas while filling compact places. Each pavement is the floor to a "room" for the pedestrians with walls, sitting places, windows to shops, and ceilings of trees or sky.

ICPI

ICPI

If the units are big, the joint lines become less prominent. When spread over a wide area like this civic space, applying a slight change in color can help define it. *ICPI*

A ribbed paver surface placed in running bond reduces drag on the tires and makes an even smoother ride for wheelchair users. *Courtesy of Nikko*

This small area presents a delightfully loose interpretation of running bond. *Courtesy of Nikko*

Widened, permeable joints infiltrate runoff from parking spaces. The tightly spaced units mark the driving area. *ICPI*

In this parking lot, both the driving and parking areas are permeable. *ICPI*

This is a concrete paver shape that bonds with its neighbors, interlocking horizontally and vertically for increased load-bearing capacity while set in running bond. *Courtesy of SF Concrete*

You can't tell the units interlock underneath just by looking at the antique street surface. *Courtesy of SF Concrete*

An innovative, cost-effective solution to a bridge deck: install, wear, and simply replace concrete pavers instead of demolishing and (re)pouring an expensive cast-in-place deck. *Courtesy of E. P. Henry Corporation*

Chapter 4: Adverbs
Herringbone

Courtesy of Barkman Concrete

Adverbs give the action words of verbs their manner, degree, and result. Herringbone patterns qualify as adverbs because of their ability to vary the manner and degree of movement. Their multi-directional rhythm can be used to vary the amount of action, movement, and direction in the pavement. This is achieved by varying unit sizes, colors, or pattern orientation.

Unit Size

To create a herringbone pattern, the length of each paver needs to be at least 1-1/2 times longer than the width. Most pavers are twice as long as wide. Occasionally, herringbone will be created with units three times long as wide. Here are some examples where the unit length is 1-1/2 times the width.

Courtesy of Barkman Concrete

Courtesy of Unit Paving

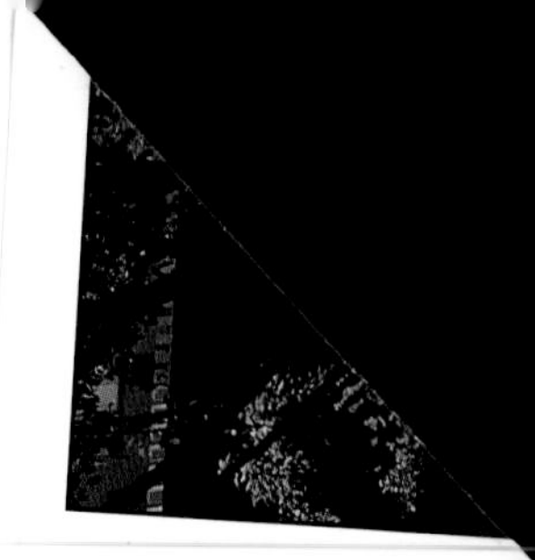

Courtesy of Pavestone

This pattern inserts an extra square paver into the herringbone to accent movement in one direction. *Courtesy of Cambridge Pavers*

Classic herringbone uses 4 x 8 inch (100 x 200mm) units, i.e., twice as long as they are wide.
Courtesy of Barkman Concrete

Courtesy of Barkman Concrete

In the right setting, herringbone pattern has a calming effect. *Courtesy of Brown's Concrete*

As the essential pattern for streets, herringbone zigs and zags, calming vehicular traffic. Gardens extend the extra front yard space. *Courtesy of David R. Smith*

Narrow pavers (three times longer than wide) create more delicate patterns. The pavers are roughened (or tumbled) during manufacture to give them an antique appearance.

Courtesy of PaverModule

Courtesy of PaverModule

While using the same size paver and similar colors, compare the texture and character of each using tumbled and non-tumbled finishes.

Courtesy of PaverModule

Courtesy of PaverModule

Pattern Orientation

Unlike stack and running bonds, herringbone patterns have no continuous joint lines. The joint lines typically change direction every 1-1/2 pavers. Depending on the view, this creates dramatic design effects. From one view, the pattern receives is visual strength from the repetition of interrupted (but implied) joint lines. From another view, the alignment of the paver surfaces, rather than joint lines, suggest direction.

Implied joint lines follow from the entrance to a building. *Courtesy of Barkman Concrete*

Here implied joint lines flow across this driveway. *Courtesy of Paver Systems*

Herringbone can spread in any direction with a consistent pattern. *Courtesy of Anchor Block Company*

ICPI

When viewed from different places, the pattern's repetition gives direction at Turner Field. *Courtesy of Site Technologies*

Cuts in the herringbone pattern create a star. *Courtesy of Site Technologies*

You can't miss it. The stack bond arrow is visible because of the contrasting color of the herringbone pattern around it. *ICPI*

Color and Graphics *From* the Pattern

A mathematical formula sets the color changes in the herringbone pattern, looking down at this bank entrance. *Courtesy of Jerry Clapsaddle*

Advancing and receding colors framed in distinct fields divide this plaza, making it more human scale. *Courtesy of Pavestone*

The color can simply be sprinkled in the pattern on streets. *ICPI*

Courtesy of Pavestone

Color can also be sprinkled on sidewalks and entrances. *Courtesy of Pavestone*

Courtesy of Pavestone

The animals at the entrance to the Johannesburg Zoo in South Africa find their form the herringbone pattern and colors.

ICPI

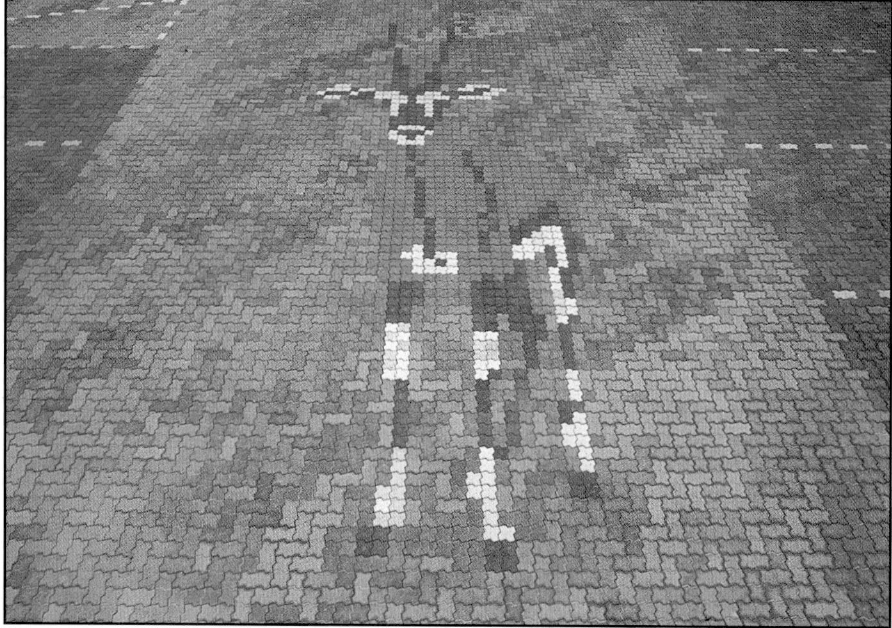

ICPI

ICPI

Filling Small Places

Like running bond, filling small places with herringbone makes surfaces feel a bit larger while adding some intimacy.

Courtesy of Mutual Materials

Courtesy of E. P. Henry Corporation

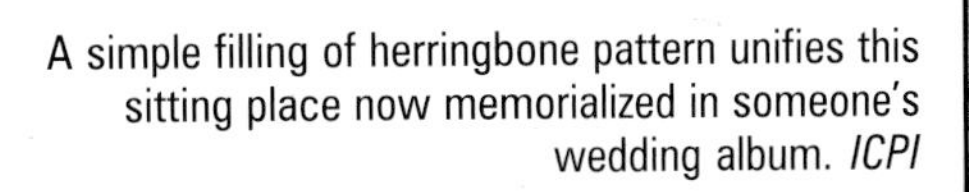

A simple filling of herringbone pattern unifies this sitting place now memorialized in someone's wedding album. *ICPI*

Workhorse Herringbone

The largest installation of concrete pavers hails at the Port of Oakland, California, with over 4 million square feet (400,000 m²) ready to support shipping containers. *ICPI*

In the world of segmental paving, the herringbone pattern is the workhorse. Essential in vehicular areas, the pattern can be placed in a 45 or 90 degree angle with respect to the direction of the traffic. Structurally, herringbone offers the greatest degree of interlock when compared to running bond, parquet, or stack bond patterns.

The beauty of industrial, port, and airport pavements with herringbone patterns goes deeper than the surface. Concrete pavers in these applications are primarily used for their engineered, functional benefits rather than for supporting design of the overall environment and place. One functional benefit is their capability to take huge loads. For example, the wheel loads of equipment in ports and airports can be roughly the same, about 50,000 lbs. (22,700 kg). These loads are five to six times higher than wheel loads from highway trucks. The high loads for port and airport pavements require careful engineering of base or foundation materials under the pavers. ICPI has design guides for street, port, and airport pavement applications.

Another plus for pavers in these applications is easy removal and reinstatement after an underground utility repair. This reduces the down time in busy ports and airports where a stationary ship or delayed aircraft only means lost income. Interlocking concrete pavement is the only one that comes with a built-in zipper. Unlike asphalt, concrete pavers are not weakened by opening and closing, and there are no ugly patches.

In many ports and some airports, the soils under them settle slowly. Unlike other pavement materials, concrete pavers allow some degree of settlement without cracking. Should settlement be excessive, the pavers and bedding sand can be removed, the top of the base raised, and the sand and pavers reinstated.

Pavers at the Port of New Orleans, Louisiana, take huge loads from stacked containers and from equipment used to move them The containers can weigh as much as 50,000 pounds (22,700 kg). *ICPI*

Concrete pavers support several cross taxiways at Dallas/Ft. Worth International Airport in Texas. *ICPI*

The largest airport installation is at Hong Kong International Airport with 4 million feet² (400,000 m²) next to the terminal building. *ICPI*

Mechanical Installation

Many street, port, and airport projects use mechanical installation. Concrete pavers are manufactured in a herringbone pattern and brought to the site stacked in layers. Specially designed machines grab and place each layer on the prepared base and sand. Some projects have used several of these special machines to speed installation. Many projects use other paver shapes and layer configurations specifically designed for placement by mechanical equipment.

ICPI

With mechanized equipment, one layer can be placed as fast as every 20 seconds on a well-organized job site. *ICPI*

Chapter 5: Informal Conversations

Random Patterns

Courtesy of E. P. Henry Corporation

Random patterns often consist of three or four units of rectangles and squares grouped together. When groupings are combined at various orientations, a random pattern appears. Like informal conversations, there is an order underlying the random appearance. It allows spontaneous exchange and an easy change in direction.

Informal conversations don't always have a specific direction. They have built-in flexibility and casualness. Likewise, random patterns don't rely on the visual strength of continuous joint lines and hence are non-directional. They fit well around the home with its less structured roles and casual conversations. They support small places where being and talking is more important than going and doing.

In today's economy, stone pavers command a price of at least four or five times that of concrete pavers. Fortunately, the beauty of stone resides in pavers through subtle color blends in the concrete. Like pre-washed jeans, concrete pavers receive surface treatments to achieve nature's aged, weathered, and relaxed look.

Driveways and Entrances

At first glance, the driveway might be a herringbone pattern. A closer look reveals a random pattern with a small hint of formality from the smooth surfaces and edges. *Courtesy of Lafarge Pavers and Walls*

This pattern suggests a random mix of stack, running, and herringbone bonds, loosely arranged and mixed with pavers of various sizes. *Courtesy of PaverModule*

Random patterns can be angled across the driveway to add informality.

Courtesy of Borgert Products

Courtesy of Pavestone

Opposite page: Stack bond from the driveway to the steps directs guests to the entrance while a random pattern fills the rest of the driveway. *Courtesy of Belgard-Permacon*

These mottled driveways took their cues from the brick walls.

Courtesy of Lafarge Pavers and Walls

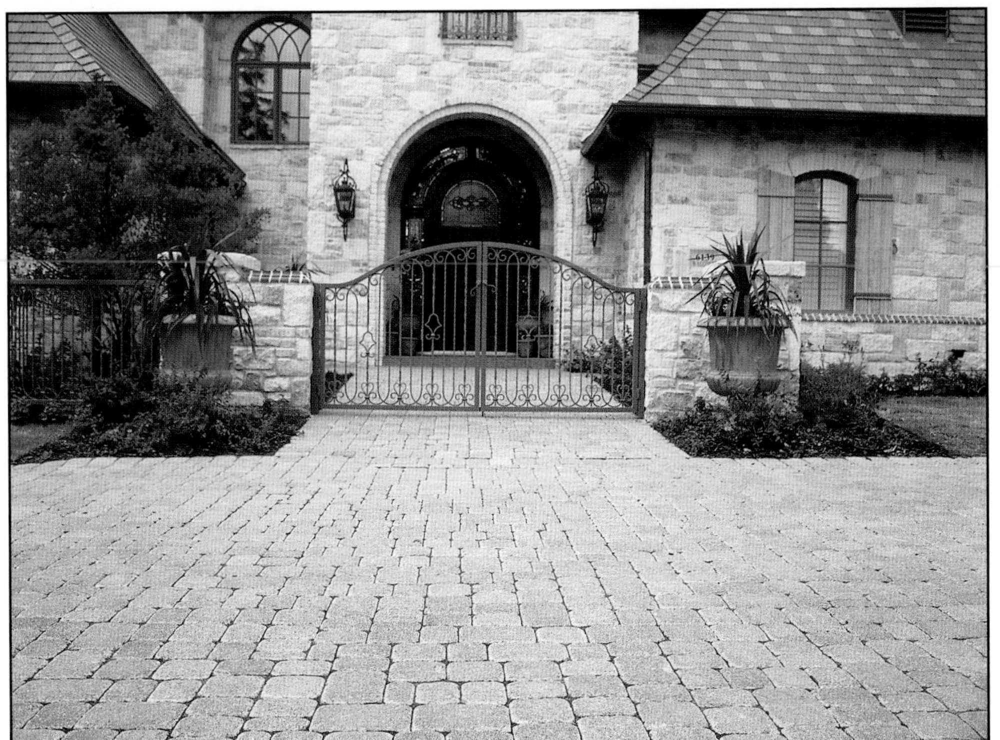

Courtesy of Pavestone

The random stone walls advised this driveway to go random. *Courtesy of Belgard-Permacon*

Larger square units make a random pattern breathe more easily. *Courtesy Belgard-Permacon*

Opposite page: Three courses along the edge bring closure to the field of random pavers. *Courtesy Belgard-Permacon*

Three courses, each with units larger than the next, add grace to ending the random field of pavers. *Courtesy of Lafarge Pavers and Walls*

The rustic nature of log homes likely inspired the use of random stone-like pavers

Courtesy of Borgert Products

Courtesy of E. P. Henry Corporation

This raised patio enlarges a backyard that's just about as wide as the house. The changes in levels add to the roominess. *Courtesy of Belgard-Permacon*

Patio chairs find their place where light and temperature are just right for people. *Courtesy of Lafarge Pavers and Walls*

Sometimes the quiet solitude of a place can say much. *Courtesy of Lafarge Pavers and Walls*

Opposite page: Hard and soft views are combined with a rock garden. *Courtesy of Borgert Products*

Soft and hard outlooks: The view from one of these patios looks at flowers, the other at rocks. Courtesy of Lafarge Pavers and Walls

Courtesy of Lafarge Pavers and Walls

Sometimes a patio is a platform to enjoy the sounds and shade from backyard trees. *Courtesy of Belgard-Permacon*

A patio can be a platform overlooking a pond. *Courtesy of Lafarge Pavers and Walls*

The smell of grilled food is often the most attractive patio feature. *Courtesy of Barkman Concrete*

Plantings extended the boundary of this small patio. *Courtesy of Pavestone*

This pattern appears almost as a running bond. A closer look, however, shows that every so often, a unit is turned 90 degrees so it engages the next course. *Courtesy of E. P. Henry Corporation*

Paver patios can simply be an excuse to support a hot tub. It can be in, on, or next to the patio.

Courtesy of Belgard-Permacon

Courtesy of Belgard-Permacon

The subtle repetition of this pool deck reads as a random pattern. *Courtesy of Paseo Stoneworks*

Like hot tubs, pools often require that the concrete paver decks be organized around them while supporting other leisure activities.

Courtesy of Lafarge Pavers and Walls

Courtesy of Belgard-Permacon

Courtesy of E. P. Henry Corporation

Courtesy of Belgard-Permacon

The colors and random pattern distinguish the lines of the parking sign for disabled people. *Courtesy of Pavestone*

People find quiet at many municipal parks like Federal Hill Park in Baltimore, Maryland.

ICPI

Federal Hill Park, Baltimore, Maryland. *ICPI*

The park contrasts its calm with a view of downtown Baltimore. *ICPI*

The random pattern at this Porsche dealership suggests that jeans may be worn during test drives. *Courtesy of Stone Products*

Chapter 6: Salutations, Discourse, and Communion

Circles and Arcs

Courtesy of E. P. Henry Corporation

Courtesy of E. P. Henry Corporation

A circular area almost always says "move closer, you're in an important place." Concrete paver circles suggest this at entrances, public plazas, and in intimate spaces. At entrances, circles issue salutations by announcing and welcoming someone. In plazas, they emphasize something important like a view, a fountain, or a monument. It is the topic of conversation and focused discourse. In small, intimate places, circles suggest drawing closer to others in communion, within a protected social space. In addition, circles sometimes define physically protected areas for things like statues, trees, and gardens.

Salutations

Circles and arcs greet and invite an approach to a home.

These circular steps reach out to visitors between two substantial walls to invite guests in. *Courtesy of Belgard-Permacon*

The doorway stoop in concrete pavers is a permanent welcome mat. *Courtesy of E. P. Henry Corporation*

Steps can reach out from the entrance to the driveway. *Courtesy of Belgard-Permacon*

Or steps can extend all the way to the sidewalk to say "come on in." *Courtesy of E. P. Henry Corporation*

Backyards can have welcome mats, too.

Courtesy of Borgert Products

Courtesy of Lafarge Pavers and Walls

Arcs in the driveway direct the viewer's eye and movement toward the home's entrance.

Courtesy of Paver Systems

Courtesy of Lafarge Pavers and Walls

19

Opposite page: This three-quarter circle directs the view from the driveway into the entrance area. *Courtesy of Belgard-Permacon*

The size, shape, colors, and placement of repeating arcs or circles affect the extent of movement. The movement can be dignified such as in this plaza with classic, regularly spaced arches. *Courtesy of Paver Systems*

A classic car and the movement of circles rarefy this driveway. *Courtesy of Lafarge Pavers and Walls*

Theres no need to display a house number to find this home. *Courtesy of Gappsi*

Circles say that now you've arrived, stop and take a look around. One way is taking a turn or two around a circular driveway.

Courtesy of E. P. Henry Corporation

Courtesy of Mutual Materials

Welcoming arcs of circles can extend past the driveway and to the development entrance. *Courtesy of PaverModule*

An uncommon mix: circular pavers in a circular driveway! *ICPI*

In public places, circles and arcs can follow the form of, and embrace, other site features such as pools, fountains, and trees in plazas and parks.

The sounds from the fountain in this small park in Saint Foy, Quebec, washes out the city noise and provides a refuge for office workers. *ICPI*

A more effective way to wash out urban sounds is to jump in the fountain. *ICPI*

The arcs gently follow the form of the swimming pool at this Miami Beach hotel. *Courtesy of PaverModule*

Courtesy of PaverModule

Circular fountains embraced by circles.

Courtesy of Mutual Materials

Courtesy of Mutual Materials

A fountain outside the Florida Aquarium in Tampa is hugged by concrete paving slabs. *ICPI*

ICPI

The rippling water softens and animates pavers in a fountain outside the Museum of Civilization in Ottawa, Ontario. *ICPI*

A circle of pavers can be the center and focus of a feature other than a fountain.

This state park looks out to the New York City skyline. *Courtesy of Argus Construction*

The big fountain in the sky was working well that day. Rain kept visitors from this Tokyo, Japan plaza and fountain. *ICPI*

Since opening, over 50 million fans have passed through Toronto's SkyDome stadium. Paver circles mark spots to admire the stadium's permanent fans. *ICPI*

Communion

Communion and intimacy begin by being close to other people or to nature. Both half circles draw near to a resting place by water provided by nature. One is at the end of a path, the other alongside.

Courtesy Borgert Products

Courtesy of Lafarge Pavers and Walls

Besides lakes and ponds, water from a backyard pool has a calming effect, especially in this sitting area under some trees. *Courtesy of Belgard-Permacon*

The circle around the whirlpool could imply a place for more private conversations. *Courtesy of Lafarge Pavers and Walls*

When nature's oceans, lakes, or ponds aren't available to a backyard, a pool—be it tiny and made with rocks, or a large swimming pool—still extends an invitation to stay, rest and chat a while.

Courtesy of Lafarge Pavers and Walls

Courtesy of Gappsi

Courtesy of Belgard-Permacon

Sometimes a small circle can border the garden, or the garden borders the circle. It provides many viewpoints. *Courtesy of E. P. Henry Corporation*

Courtesy of Gappsi

Enclosed by trees, this small meditation area hidden in a public park is a small maze to on which to walk slowly. *Courtesy of Paveloc and Artistic Pavers*

Courtesy of Paveloc and Artistic Pavers

Concrete pavers can be tree huggers.

ICPI

ICPI

ICPI

Chapter 7: Changing the Subject

Steps and Edges

ICPI

Courtesy of E. P. Henry Corporation

Introducing and closing conversations make communication memorable. Such transitions range between rude or polite, humorous or sad, sudden or gradual. Steps provide transitions from one level to another. Edges slow down, close, and stop pavement patterns. Steps and edges both offer transitions from one place to another, transitioning moods, feelings, and character.

Steps

Front steps provide a physical and mental transition from the street and driveway to the door. Depending on the slope and distance to the door, the transition by steps can be very deliberate and grand, as shown below, or separated into smaller stages as shown below and on the opposite page.

Courtesy of Lafarge Pavers and Walls

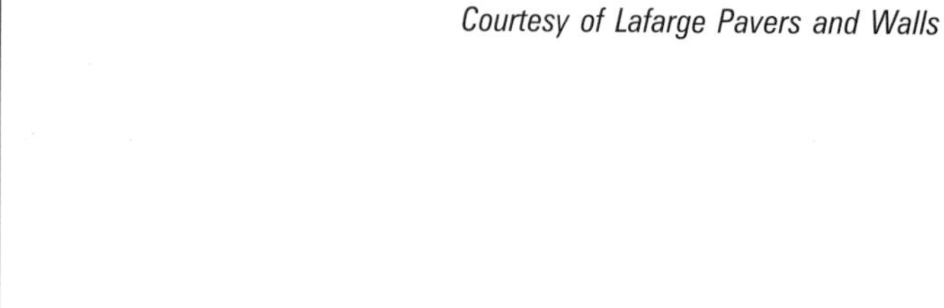

Courtesy of Belgard-Permacon

Courtesy of Belgard-Permacon

Courtesy of E. P. Henry Corporation

Steps and planters combine and turn a corner to conserve space. *Courtesy of Belgard-Permacon*

When the distance from driveway to entrance is short and flat, a simple transition may be all that is needed. *Courtesy of PaverModule*

Level changes in backyards don't require the same grand transition in levels as some front yards.

These steps turn back toward the planter. *Courtesy of Mutual Materials*

Sometimes a small bend in the step can soften the transition. *Courtesy of Belgard-Permacon*

Steps wrap around this wood deck to the lower elevations of the grade. *Courtesy RM Stonescaping*

Steps from an entrance onto a patio should be minimized so that usable patio space is maximized. *Courtesy of E. P. Henry Corporation*

Steps transform and enlarge into stage areas.

Courtesy of E. P. Henry Corporation

Courtesy of E. P. Henry Corporation

Courtesy of E. P. Henry Corporation

The short transition from patio to yard is an opportunity to express informality that can contribute to a relaxed setting.

Courtesy of Belgard-Permacon

Courtesy of E. P. Henry Corporation

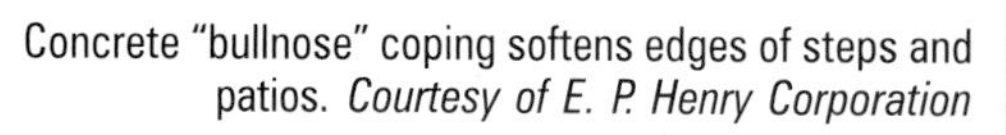

Concrete "bullnose" coping softens edges of steps and patios. *Courtesy of E. P. Henry Corporation*

Edges

Like ending conversation, ending pavement can be abrupt and hard, gentle and soft, or between these extremes. Hard edges often involve another material or wall to stop the pavement.

A hard edge is when the pavement stops against another material such as a stone or concrete curb. *ICPI*

Sometimes the edge can be hard, but not look so hard, due to its color, texture, and scale like the edge here. *ICPI*

Transitioning from one pattern to the next means stopping a pattern, providing a short break with another pattern, and starting the other. *Courtesy of Pave Tech*

Ordinary manholes demonstrate how different sized units may be needed to transition from one area to the next.

ICPI

ICPI

A concrete garden wall stops the pavers. The small units between the wall and gutter cushion the visual impact and fill a small space that can accommodate larger units. *ICPI*

Paver courses (i.e., soldier, sailor, etc.) typically define edges. Examples abound throughout this book. Mixing materials at the edge, however, is unusual. This example combines asphalt, concrete and stone pavers achieving contrast through color, pattern, and texture. *ICPI*

A softer edge is created when the pavement simply stops at the grass. In this case, the grass is drifting into the pattern to further soften it. *ICPI*

Grid pavers make the transition from hard, solid concrete paver to grass. The dark color of the grids helps define the yard space adjacent to apartments. *ICPI*

Vegetation can drift over the pavement and soften its edges as well. *Courtesy of Pavestone*

A slight drop in elevation makes a hard edge, but there's a soft landing here. *Courtesy of Belgard-Permacon*

Edges can have surprises. *Courtesy of Lafarge Pavers and Walls*

The softest edge is water.

ICPI

ICPI

Special Shapes and Patterns

ICPI

Like vocabulary and language, design of concrete paver shapes and finishes is constantly evolving. Trends in culture, fashion, and technology bring new words, phrases, and meanings. Such invention can result in a fresh point of view.

At least two hundred shapes exist in concrete pavers. This chapter presents some shapes and patterns that aren't the typical square or rectangle. They're different approaches, offering greater variety to the vocabulary and diction of segmental paving. Like new, unusual, or seldom used words, the challenge with clever paver shapes and patterns is finding the right fit for projects, i.e., the right sentence, paragraph, story, and site for them. When found, they can bring fresh identity and meaning to the message of a place.

Certain shapes, like those here and on the opposite page, can be placed in running bond or herringbone and each will generate the same pattern. Each unit consists of an octagon joined to a square.

The neighborhood sidewalk, an important stage for weaving the fabric of daily social interaction, gets special pattern treatment. *Courtesy of E. P. Henry Corporation*

ICPI

A sister pattern: the octagon and square are made as separate units and colors, then placed in this striking, tile-like pattern. *Courtesy of PaverModule*

Messages can be from the pattern or in the pavers themselves. Monument engraving equipment cuts out the words in pavers. They're great fundraisers for municipal projects and public spaces. Practically any shape can be engraved.

The sale of engraved pavers to rock fans enabled a long-term source of funds for maintenance of the plaza space around Cleveland's Rock and Roll Hall of Fame and Museum in Ohio. *ICPI*

The Winnipeg Zoo in Manitoba raised funds for an expansion project with engraved pavers. *Courtesy of Barkman Concrete*

Opposite page: Even the animals got involved in making an impression on the project! *Courtesy of Barkman Concrete*

Concrete pavers can be molded into practically any form. Flowing shapes like this one are an alternative to the sharp corners of geometric shapes. *Courtesy of Cambridge Pavers*

This small passage between two buildings holds a unique pattern of fish-like shapes. Could the pattern transform into an Escher print? *Courtesy of David R. Smith*

Sharp corners of square units are rounded out with round lights of different colored glass. An entire field or grid of lights could present a fascinating show. *ICPI*

Concrete pavers jump from the surface with three-dimensional designs. They are created with a simple combination of shapes and colors.

ICPI

ICPI

ICPI

ICPI

Stars and stripes forever.

ICPI

ICPI

ICPI

Visual smorgasbords: Like carpet or cabinet samples, some paver manufacturers display products for viewing by customers. The combination of pavers presents its own set of design ideas.

ICPI

ICPI

ICPI

The family of precast paving products includes concrete grid pavers. They have a unique role in decreasing stormwater runoff by increasing infiltration. Grid pavers are often called 'turf stone' or 'grass pavers' because they support occasional vehicles while allowing grass to grow. Visually, their surface softens the impact of hard concrete. Environmentally, their use demonstrates greater sensitivity to design with nature's processes.

ICPI

Courtesy of E. P. Henry Corporation

A tree in Auckland, New Zealand receives needed air and rain to its roots from concrete grid pavers while dividing traffic at an intersection. *ICPI*

Permeable interlocking concrete pavers answered the need for a surface that could take a greater amount of vehicles than grid pavers and still reduce stormwater runoff. Permeable units are placed over an open-graded stone base that holds infiltrated rainfall. This environmentally friendly paving is seeing increased use in residential and commercial applications. While some designs can support grass, they receive more runoff when the joints are filled with small-sized aggregate.

ICPI

ICPI

ICPI

Some variations on the theme of permeable pavers.

ICPI

ICPI

ICPI

Paver Supergraphics

© Gary Knight: Gary Knight + Associates, Inc. *ICPI*

Language is packed with pictures and symbols. Their meaning reverberates from idioms, metaphors, figures of speech, and allegories from verbal, written, and visual outlets. There are two approaches to translating such signs and symbols into concrete pavers. First, shapes, colors, and patterns can be used as the framework for image making. Second, different colored units can be cut to create pictures.

From a construction point of view, the former approach requires more discipline and less labor-intensive cutting, the latter needs far less attention to the pattern and much more cutting to make all the paint-by-number pieces fit together. The former approach uses concrete pavers as the canvas, paint, and picture. The latter uses pavers for canvas and paint only. Each produces supergraphics consisting of mural-sized pictures, maps, and mosaics.

Downtown Daytona Beach, Florida, has a stream of water running through it to help keep it cool. *Courtesy of Paver Systems*

Pictures, Signs, and Symbols

The stream waters a flower-shaped plaza. *Courtesy of Paver Systems*

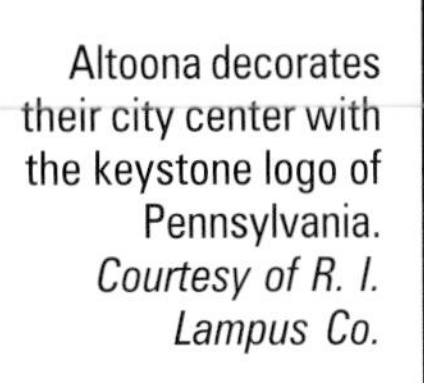

Altoona decorates their city center with the keystone logo of Pennsylvania. *Courtesy of R. I. Lampus Co.*

Courtesy of R. I. Lampus Co.

A hotel entrance is graced by giant leaves. *Courtesy of Surebond*

A flower blooms amidst a field of pavers. *Courtesy of Gappsi*

Courtesy of Surebond

A famous mouse lives in someone's backyard patio. As the kids grow, the image can be transformed to a superhero. *Courtesy of E. P. Henry Corporation*

The construction is sort of a paint-by-pavers exercise. *Courtesy of E. P. Henry Corporation*

A park in Chattanooga, Tennessee, commemorates a local Medal of Honor winner. The star design is taken from the medal. The center contains a monument to the hero.

ICPI

Courtesy of Bosse Concrete

The sidewalks of downtown Miami, Florida, are painted with bold strokes of color designed by the Brazilian landscape architect, Burle Marx. *Courtesy of PaverModule*

Courtesy of PaverModule

No, it's not painted. This coat of arms for Washington and Lee University in Virginia is delicately cut into the concrete pavers.
Courtesy of Paveloc and Artistic Pavers

Courtesy of Paveloc and Artistic Pavers

Courtesy of Paveloc and Artistic Pavers

Some examples of animals emerging from concrete pavers.

Courtesy of Pavestone

Courtesy of Pavestone

ICPI

A giant bike sprocket at Virginia Beach suggests a place for cyclists to hang out. *Courtesy of Interlock Paving Systems*

Along Virginia Beach, pavers mark a sandcastle that can't be washed away by ocean waves. *Courtesy of Interlock Paving Systems*

Near the same location, blue pavers on the walks suggest the water close by. *Courtesy of Interlock Paving Systems*

A pitch from this giant baseball helps create a walk among baseball fields.
Courtesy of PaverModule

A nearby compass is often handy for getting your bearings.

Courtesy of Gappsi

Courtesy of PaverModule

Courtesy of Interlock Paving Systems

As picture and map, the baseball diamond outside the Edison Field Stadium in Anaheim, California, provides a place for games and contests before and after the one held inside the stadium. *Courtesy of Angelus Block Co.*

Clock watching must be a major pastime by those in the offices that surround this sundial. *ICPI*

Those who win and finish the Vancouver Grand Prix in British Columbia start their celebration here. *ICPI*

A mosaic of pavers can make waves in some projects.

Courtesy of Tremron

ICPI

Everywhere you go in the world, there will be concrete pavers. *Courtesy of Game Time*

Courtesy of Lafarge Pavers and Walls

African Kente cloth patterns in this small park in Miami, Florida, are welcomed into the urban fabric.

© Gary Knight: Gary Knight + Associates, Inc.

Make your own art; speak your own language: concrete unit paving as canvas and platform. *ICPI*

Chapter 10: Vocabulary Builders

A Resouce Guide

Every language has regional dialect, accents that contribute to local color and character. The same is true with concrete pavers. The popularity of various shapes and colors—and their availability—varies regionally across North America. The companies listed below would be pleased to provide additional information on shapes, patterns, colors, and finishes made in their market region. All manufacturers are members of the Interlocking Concrete Pavement Institute. Most contractors are members and are ICPI certified concrete paver installers. Manufacturers can refer experienced, ICPI certified contractors to assist with the design and construction of residential and commercial projects.

Anchor Block Company
2300 McKnight Road
North St. Paul, MN 55109
651-777-8321/Fax: 651-777-0169
jhogan@anchorblock.com
www.anchorblock.com
Based in Minneapolis/St. Paul, Minnesota, Anchor Block Company manufacturers a diversified range of concrete products including concrete blocks, licensed Anchor Retaining Wall Systems, and concrete pavers.

Angelus Block Co., Inc.
3435 Riverside Avenue
Rialto, CA 92316
909-328-9115/Fax: 909-321-0115
pavers@angelusblock.com
www.angelusblock.com
Founded in 1946, Angelus Block Co., Inc. produces concrete masonry units and interlocking concrete pavers in southern California.

Argus Construction & Equipment Co., Inc.
340 Orient Way
P.O. Box 535
Lyndhurst, NJ 07071
973-661-2626
argusvap@aol.com
Argus builds interlocking concrete pavements and is an ICPI certified paver installation contractor.

Artistic Pavers
800 Big Rock Avenue
Plano, IL 60545
630-552-3408/Fax: 630-552-1737
artpaver@aol.com
www.artpaver.com
Owner Marty Kermeen has been sculpting pavers by hand for 15 years. Specializing in labyrinth construction, he has been commissioned by the Museum of New Mexico, the University of Texas Medical Center, St. Johns Cathedral, and many others.

Barkman Concrete Ltd.
909 Gateway Road
Winnipeg, MB R2K 3L1 Canada
204-667-3310/Fax: 204-663-4854
wpgsales@barkmanconcrete.com
www.barkmanconcrete.com
Founded in 1948, Barkman Concrete manufactures precast concrete products for residential, commercial, and industrial projects. They specialize in concrete pavers, retaining walls, and site furnishings.

Basalite Concrete Products
11888 West Linne Road
Tracy, CA 95376
209-833-3670/Fax: 209-833-6039
bruce.camper@paccoast.com
www.basalitepavers.com
Basalite Concrete Products provides a wide range of architectural concrete products in the western United States. They manufacture paving stones, Keystone™ segmental concrete retaining walls, concrete block, and sack products for architects, engineers, developers, and dealers.

Bayer Corporation
100 Bayer Road
Pittsburgh, PA 15205
412-777-2000/ Fax: 412-777-7626
paul.croushore.b@bayer.com
www.bayferrox.com
Bayer manufactures Bayferrox iron oxide pigments. When mixed with concrete they make colorful concrete pavers, segmental retaining walls, and many other attractive concrete products.

Belgard-Permacon
Oldcastle
375 Northridge Road Suite 250
Atlanta, GA 30350
770-804-3363/Fax: 770-804-3369
johnkemper@oldcastleapg.com
www.oldcastleapg.com
The Oldcastle Architectural Products Group provides a wide range of architectural masonry and ornamental concrete products, roof tiles, segmental retaining walls, and interlocking pavers to professional builders, contractors, dealers, engineers, and architects.

Borgert Products, Inc.
P.O. Box 39
St. Joseph, MN 56374
320-363-4671/Fax: 320-363-8516
www.borgertproducts.com
Established in 1923, Borgert Products manufactures UNI® interlocking concrete paving stones, the MESA® retaining wall system, and decorative paving slabs.

Bosse Concrete Products Company
7414 Jonesboro Road
Jonesboro, GA 30236
Dolph.bosse@oldcastleapg.com
www.belgardhardscapes.com
770-478-08817/Fax: 770-471-2128
Bosse Concrete Products Company is an Oldcastle Company located in Jonesboro since 1954. Bosse manufactures interlocking concrete pavers, retaining wall units and patio products.

Brown's Concrete Products, Ltd.
3075 Herold Drive
Sudbury, ON L1M 1B5 Canada
705-522-8220/Fax: 705-522-2732
mkhbrown@isys.ca
Manufacturing in Sudbury since 1907, Brown's offers a large selection of concrete pavers, segmental concrete retaining walls, and architectural masonry for residential, commercial, and industrial applications.

Cambridge Pavers Incorporated
Jerome Avenue
P.O. Box 157
Lyndhurst, NJ 07071
201-933-5000/Fax: 201-933-5532
cambridge@cambridgepavers.com
www.cambridgepavers.com
Cambridge Pavers manufactures Cambridge Pavingstones with Armortech and segmental retaining walls. Distributors of Cambridge products are located from the New England states through North Carolina.

Davis Colors
3700 East Olympic Blvd.
Los Angeles, CA 90023
323-269-7311/Fax: 323-269-1053
info@daviscolors.com
www.daviscolors.com
Davis Colors manufactures powder, liquid and granular low-dust color pigment additives and automatic dosing systems used by manufacturers of concrete pavers and masonry products. Now a brand of Rockwood Pigments, Inc., Davis Colors has served the U.S. concrete industry since 1952.

E. P. Henry Corporation
P.O. Box 615
201 Park Avenue
Woodbury, NJ 08096
856-845-6200/Fax: 856-845-0023
info@ephenry.com
www.ephenry.com
E. P. Henry is a manufacturer of architectural concrete masonry and hardscaping products including interlocking concrete pavers, segmental retaining walls, garden walls, and patio products.

Gappsi Inc.
311 Veterans Memorial Highway
Commack, NY 11725
631-543-1177/Fax: 631-543-1188
info@gappsi.com
www.gapsi.com
Gappsi Inc. is a paving stone design and installation company. The owner of the company, Giuseppe Abbrancati, is the licensor of several decorative and versatile paving stone shapes. He has designed and his company distributes flexible edge restraint systems for paving stones.

Jerry Clapsaddle
2215 Martha's Road
Alexandria, VA 22307
703-765-3619/Fax: 703-765-3036
jerryclapsaddle@earthlink.net
Jerry Clapsaddle is an Associate Professor of Art and Visual Technology at George Mason University in Fairfax, Virginia. He has been creating public art with segmental paving since 1985.

Interlock Paving Systems, Inc.
802 W. Pembroke Avenue
Hampton, VA 23669
757-723-0774/Fax: 757-723-8895
info@interlockonline.com
www.interlockonline.com
Interlock Paving Systems began manufacturing concrete pavers in July 1983. They offer a comprehensive line of paving products in 36 shapes and 21 standard colors.

Interlocking Concrete Pavement Institute
1444 I Street, N.W. – Suite 700
Washington, DC 20005-6542
202-712-9036/Fax: 202-408-0285
ICPI@ICPI.org
www.icpi.org
Founded in 1993, the Interlocking Concrete Pavement Institute represents member producers, contractors, and suppliers to the segmental concrete paving industry in the United States and Canada. The ICPI publishes resources for homeowners, architects, engineers, and landscape architects on the design, specification, construction, and maintenance of interlocking concrete pavers, concrete paving slabs, concrete grid pavers, and permeable interlocking concrete pavements.

Lafarge Pavers and Walls
51744 Pontiac Trail
Wixom, MI 48393-1906
248-684-5004/Fax: 248-684-2726
rick.stinchcombe@lafargecorp.com
www.lafargepavers.com
Lafarge is one of the largest building materials manufacturers in the world. Products include interlocking pavers, segmental concrete retaining walls, other concrete products, aggregate, cement, and ready-mix concrete.

Matt Stone, Inc.
P.O. Box 1929
Zephyrhills, FL 33539
813-783-1970/Fax: 813-783-2728
info@mattstone.com
www.mattstone.com
Matt Stone manufactures and sells concrete landscaping products throughout the southeastern United States from facilities in Florida, Georgia, South Carolina, Kentucky, and Alabama.

Mutual Materials Co.
P.O. Box 2009 605-119th Avenue NE
Bellevue, WA 98009
425-452-2300/Fax: 425-637-0771
info@mutualmaterials.com
www.mutualmaterials.com
Since 1900, Mutual Materials has been a masonry products manufacturer in the Pacific Northwest. They manufacture concrete pavers, segmental retaining walls, brick, concrete block, and DesignMix mortars and grouts.

Nihon Kogyo Co. Ltd. (Nikko)
721-2 Kami-Fukuokacho
Takamatsu 760-0077 Japan
+81-87-831-2828/Fax: +81-87-861-5310
wadasyohei@ms.nihon-kogyo.co.jp
www.nihon-kogyo.co.jp
Nikko manufactures concrete pavers, permeable pavers, architectural concrete masonry units, and other precast concrete products in Japan.

Paseo Stoneworks & Design
123 S. San Francisco Street, #1
Flagstaff, AZ 86001
520-774-6949/Fax: 520-774-8061
paseostoneworks@excite.com
www.paseostoneworks.com
Paseo Stoneworks & Design strives for perfection in the design and construction of concrete paver installations in northern Arizona. Paseo's team includes a professional estimator and design/layout staff to ensure high quality projects.

Paveloc Industries, Inc.
8302 South Route 23
Marengo, IL 60152
815-568-4700/Fax: 815-568-1210
mail@paveloc.com
www.paveloc.com
Paveloc Industries, Inc. is a manufacturer of interlocking concrete paving stones and Crete™ Stone retaining walls.

Pave Tech, Inc.
P.O. Box 576
Prior Lake, MN 55372
952-226-6400/Fax: 952-226-6406
sales@paveedge.com
www.pavetech.com
PAVE TECH is a leader in supplying PAVE EDGE, PROBST tools and handling equipment, PAVE CHEM adhesives, cleaners, sealers, SandLOCK Joint Sand Stabilizer, DeepRoot, and Enviropave.

PaverModule, Incorporated
1590 North Andrews Avenue Extension
Pompano Beach, FL 33069
954-972-7400/Fax: 954-972-7433
www.PaverModule.com
PaverModule is a manufacturer of interlocking concrete pavers in Florida and Nevada specializing in unique colors, patterns, and textures.

Pavestone Company
4835 LBJ Freeway – Suite 700
Dallas, TX 75244
972-404-0400/Fax: 972-404-4379
bobby@pavestone.com
www.pavestone.com
Pavestone manufactures and distributes interlocking concrete pavers, segmental concrete retaining walls, hardscape systems, and architectural precast concrete products in 40 states from 16 production facilities coast to coast.

Paver Systems
7165 Interpace Road
West Palm Beach, FL 33407
561-844-5202/Fax: 561-844-5454
garyr@tarmacamerica.com
www.tarmacamerica.com
Paver Systems has been in Florida since 1974 with three plants in West Palm Beach, Orlando and Tampa. As a member of the ICPI, UNI-Group, and Symrah Licensing, Inc., we offer a large assortment of shapes and colors to commercial and residential markets.

Perfect Pavers, Inc.
528 N.W. 1st Avenue
Ft. Lauderdale, FL 33301
954-779-1188/Fax: 954-779-1049
msperfect@mindspring.com
Perfect Pavers is an interlocking paver installation company in its 15th year of serving residential, condominium, commercial, and pool builders in Florida and the Bahamas.

R. I. Lampus Company
316 R. I. Lampus Avenue
Springdale, PA 15144
412-362-3800/Fax: 724-274-2181
rilampus@lampus.com
www.lampus.com
Founded in 1924, the R. I. Lampus Company manufactures concrete blocks, interlocking concrete pavers, and segmental concrete retaining walls. Markets served include western Pennsylvania, eastern Ohio, and West Virginia.

RM Stonescaping Corporation
5740 Columbia Road
Medina, OH 44256
330-723-7347/Fax: 330-723-7448
rmann2@neo.rr.com
RM Stonescaping is a specialty contractor in the design and installation of residential retaining walls and interlocking concrete pavements.

SF Concrete Technology, Inc.
2155 Dunwin Drive, Suite25
Mississauga, ON L5L 4M1 Canada
piro@sfconcrete.on.ca
Licensors of SF products including SF-SLOPE-35, SF-MATORO interlocking pavers, SF-RIMA/SF-ECO permeable concrete pavers and TEGULA-TEC with horizontal and vertical interlock.

Site Technologies, Inc.
5090 Old Ellis Pointe
Roswell, GA 30076
770-993-4344/Fax: 770-587-3042
beckman@sitetechnologies.com
www.sitetechnologies.com
Site Technologies is a commercial decorative paving and hardscape contractor serving the southeastern United States.

Stone Products, Inc.
P.O. Box 110
Maple Valley, WA 98038
425-432-8696
paverproducts@home.com
Stone Products, Inc. is a specialty contractor for interlocking pavers, pedestal-set slabs, and retaining wall systems. We assist with design ideas in both residential and commercial projects.

Surebond California, Incorporated
228 Las Flores
Aliso Viejo, CA 92656
949-360-4446/Fax: 949-360-5999
surebondcalifornia@msn.com
www.surebond.com
Surebond supplies adhesives, liquid polymer sealers, and joint sand stabilization materials for interlocking concrete pavements.

Tremron
11321 NW 138 Street
Miami, FL 33178
305-825-9000/Fax: 305-823-6614
tremron01@aol.com
www.tremron.com
Tremron has manufacturing plants in Miami and Jacksonville, Florida. They produce concrete pavers, segmental retaining walls, and other concrete products.

Unit Paving, Inc.
119 Dogwood Trail Lane
Ft. Mill, SC 29715
803-802-3770/Fax: 803-802-3803
pavers@unit paving.com
www.unit paving.com
Unit Paving, Inc. specializes in residential and commercial installation of brick pavers supplying a wide variety of colors, textures, and sizes to fit your needs.

Photo Credits

Lafarge Pavers and Walls - Images, 349 Maitland Street, London, Ontario, N6B 2Y7 Canada, 519-673-1174/Fax: 519-673-4866, karsten@images.on.ca.

R. I. Lampus Company - Alexander Patho Photography, 209 North Rose Avenue, Glenshaw, PA 15116, 412-486-1621.

PaverModule - Greg Wilson Studios, P.O. Box 23505 Sarasota, FL 34277, 941-366-1212, www.gregwilsonphoto.com.

E. P. Henry Corporation - Photography by Jay Baccile, 327 King Street, Woodbury, NJ 08096, 856-845-9074 and Lawrence S. Williams, Inc., P.O. Box 694, Kimberton, PA 19442, 610-983-0227.